My first

Coloring pages.

Number tracing practice.

Problem solving.

math book.

About the author

I am the proud mother of twin boys Allen and William. At three years of age, they were diagnosed with autism, neurofibromatosis and an intellectual disability. Because of their disability and health conditions, they need individualized attention when it comes to their education. That's why I chose homeschooling. Through our homeschool journey, I have come to realize that they don't learn in the same way that other kids do. Therefore, I've had to come up with different teaching strategies to help them understand and process the world around them.

Growing up, I always enjoyed teaching others. At a very young age, I was working at a bookstore, and later on, I had the opportunity to work as a call center representative while studying biology at the University of Puerto Rico. A few years later, I decided that I wanted to become a pastry chef, so I enrolled in college one more time, gained experience doing this as a side hustle, and I was able to work as an Assistant Pastry Chef. When it was time for a new adventure, I left my native Puerto Rico and moved to the United States to become a flight attendant. This gave me the chance to be exposed to different cultures and to learn from people from all around the world.

Today, I am a full time homeschooling single-mom. Without a doubt, this has been the toughest, and yet, most rewarding job I've ever had.

About the book

When it comes to learning, we are all wired differently. It doesn't matter if we have a diagnosis of some sort, or a disability, or even if we are neurotypical. We all learn in different ways. Some people are more visual, some people are more hands-on, some learn through music, some learn by repetition, and the list goes on and on.

As a homeschooling mom, my goal is to share with the world the strategies that are working for us, so that anybody that has a child that struggles with learning a specific skill, can try them. I aim to tackle each skill with a different approach than the traditional one.

The purpose of this book is to introduce kids to basic math skills, such as number recognition, counting, and problem solving. While doing so, kids can also learn other skills. These skills include coloring, hand-eye coordination, fine motor skills and tracing. It's also a great opportunity to practice colors and everyday objects.

Coloring pages

zero

1

one

2

two

3

three

4

four

5

five

6

six

7

seven

8

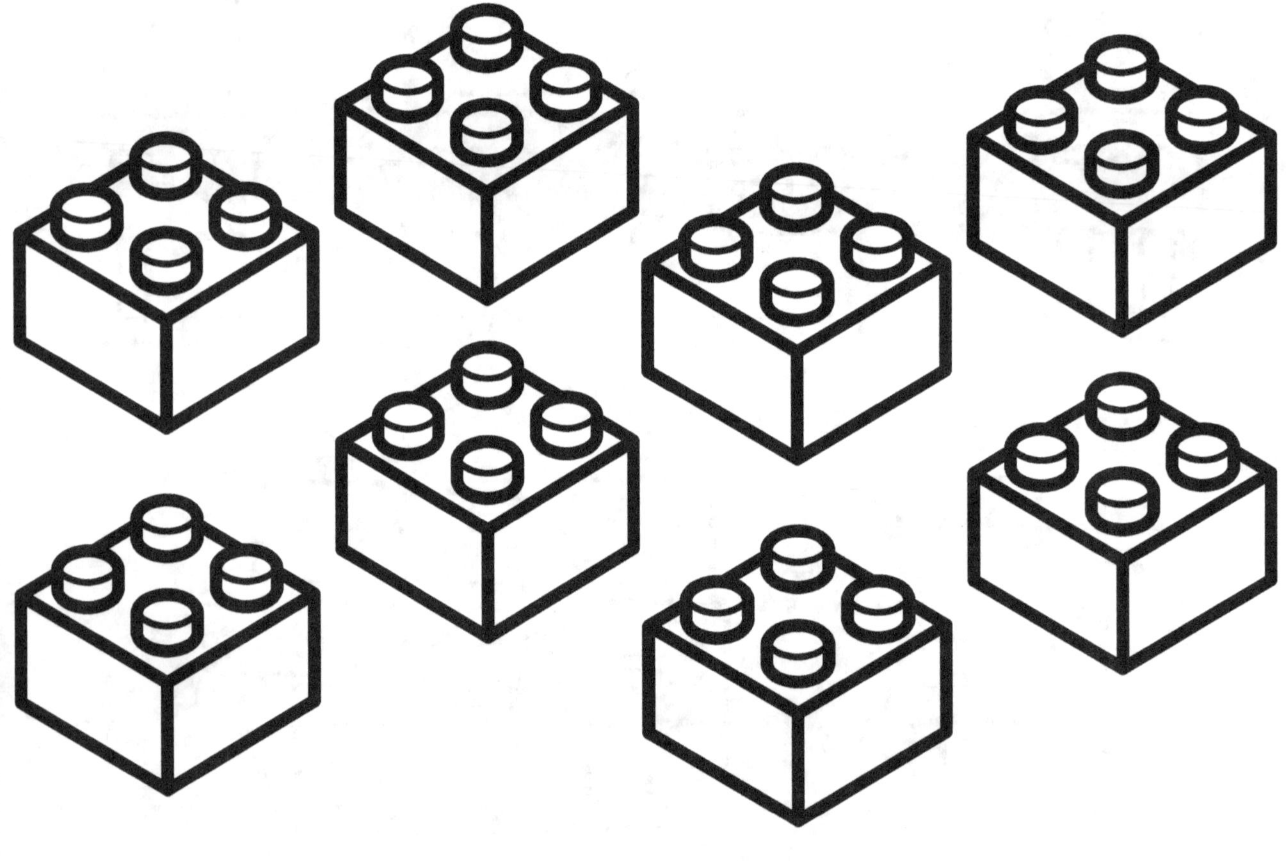

eight

nine

10

ten

Dot marker

coloring

pages

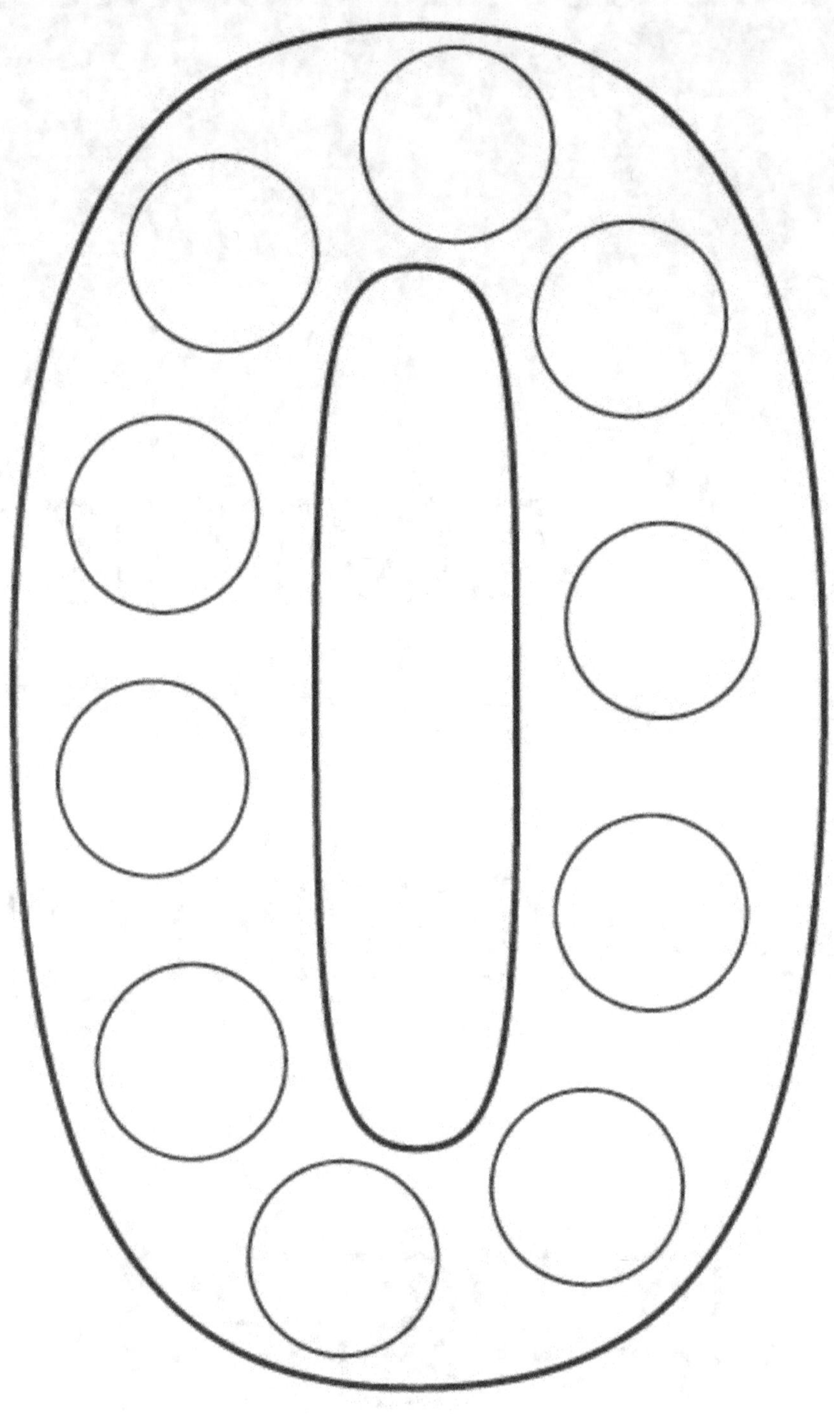

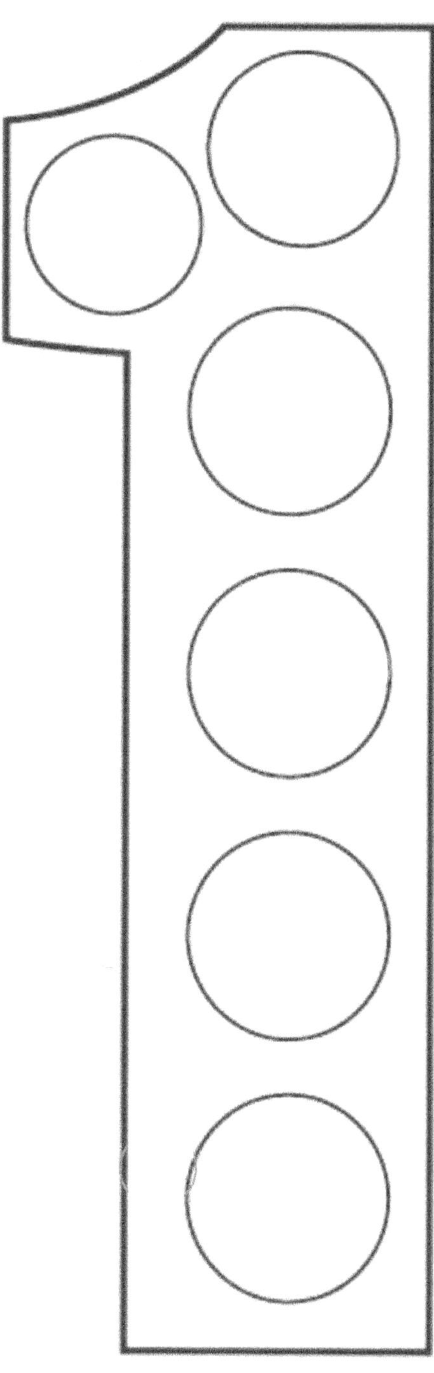

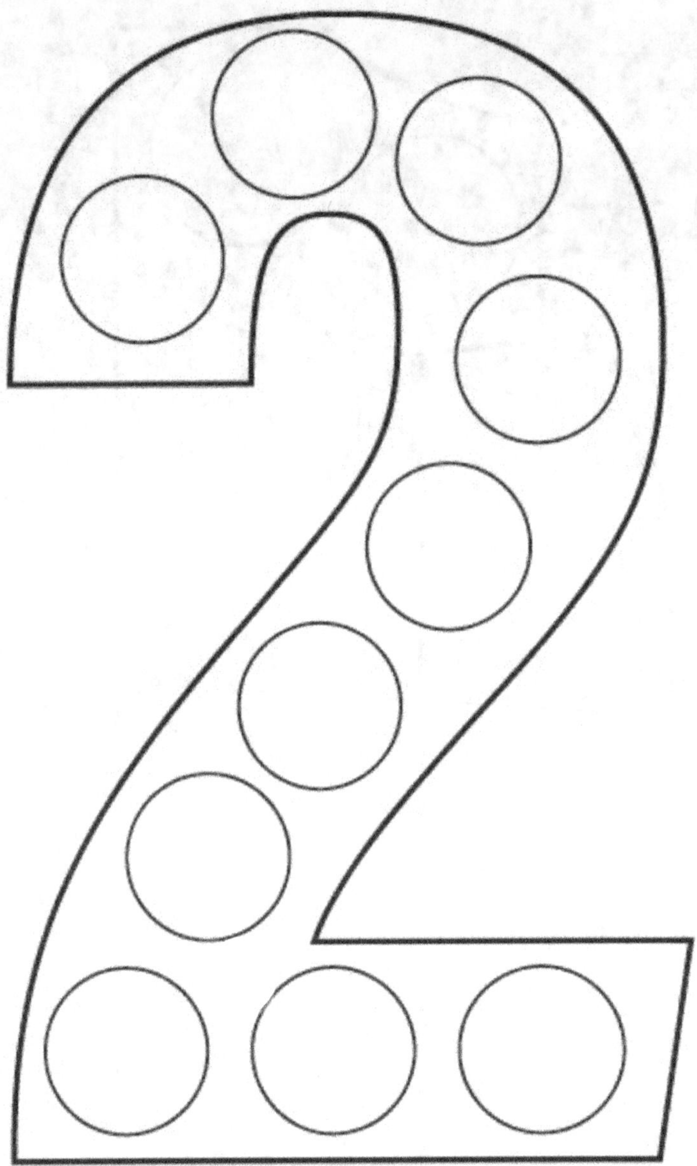

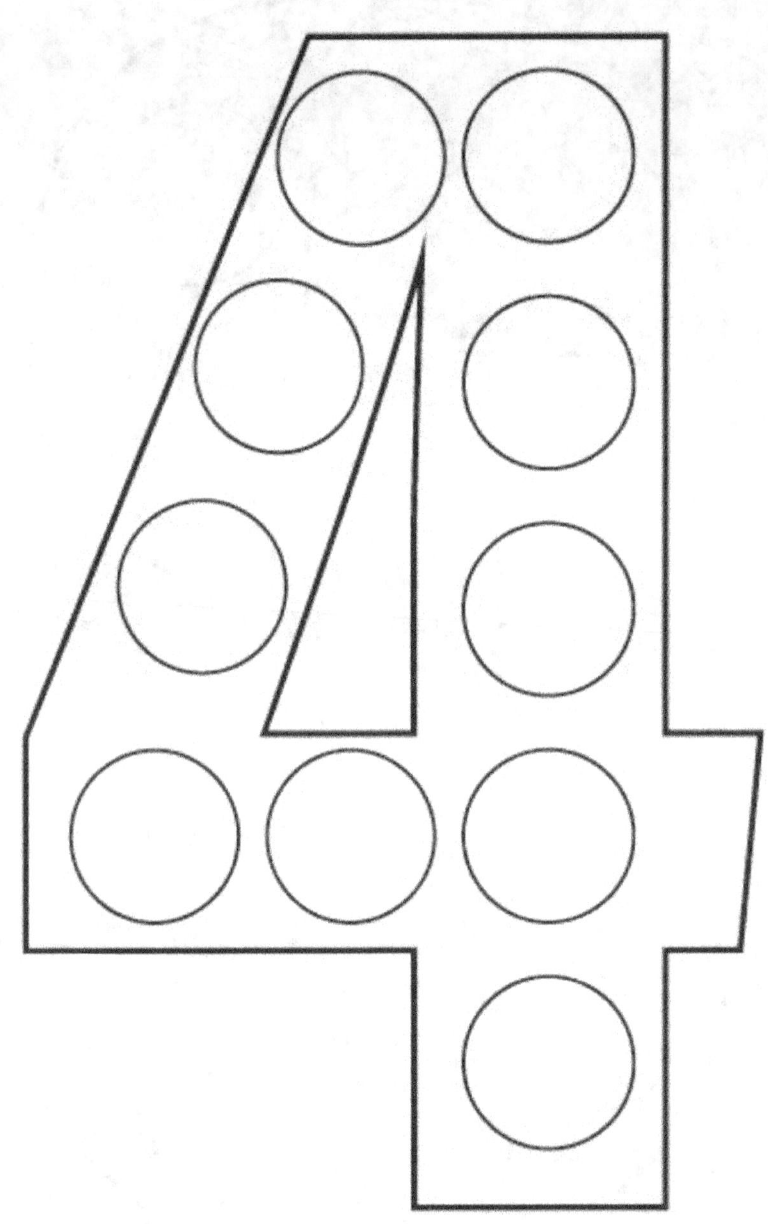

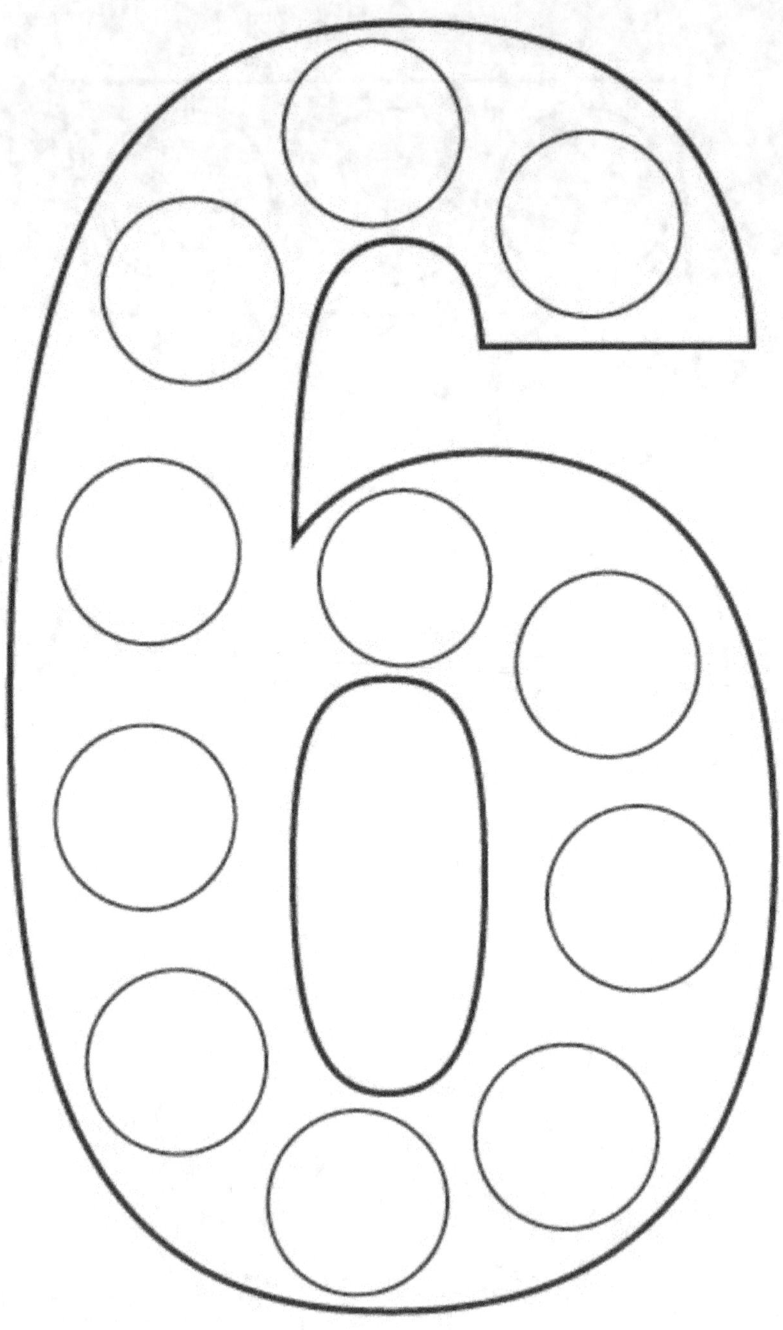

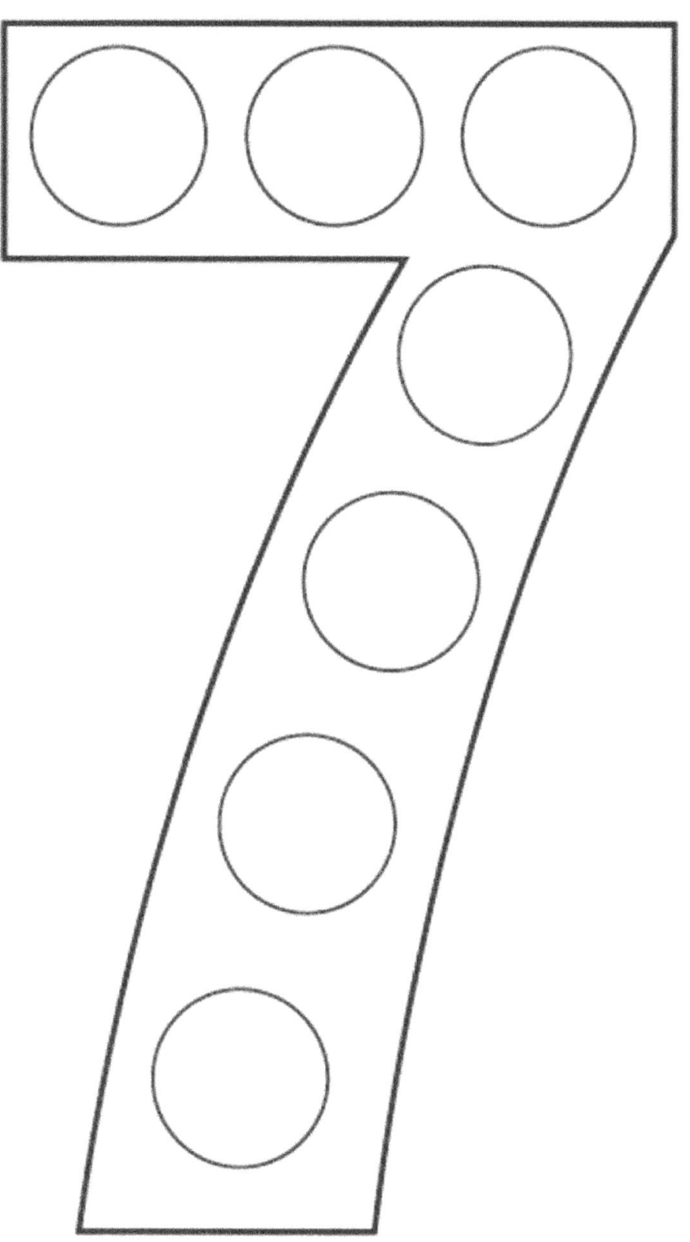

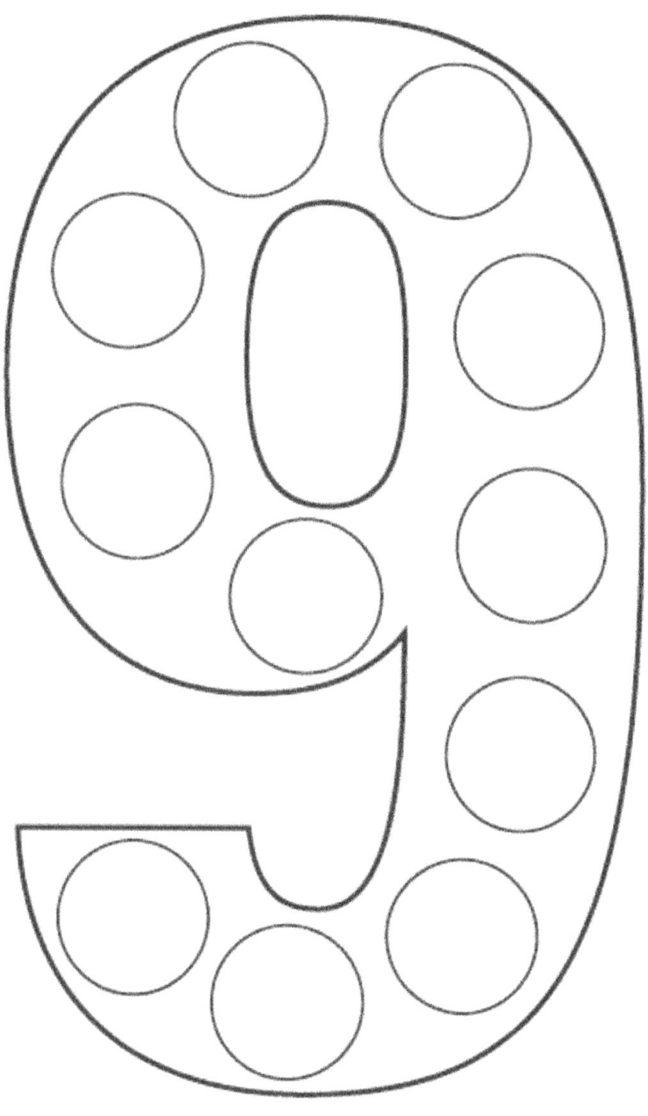

Number tracing practice

0

zero

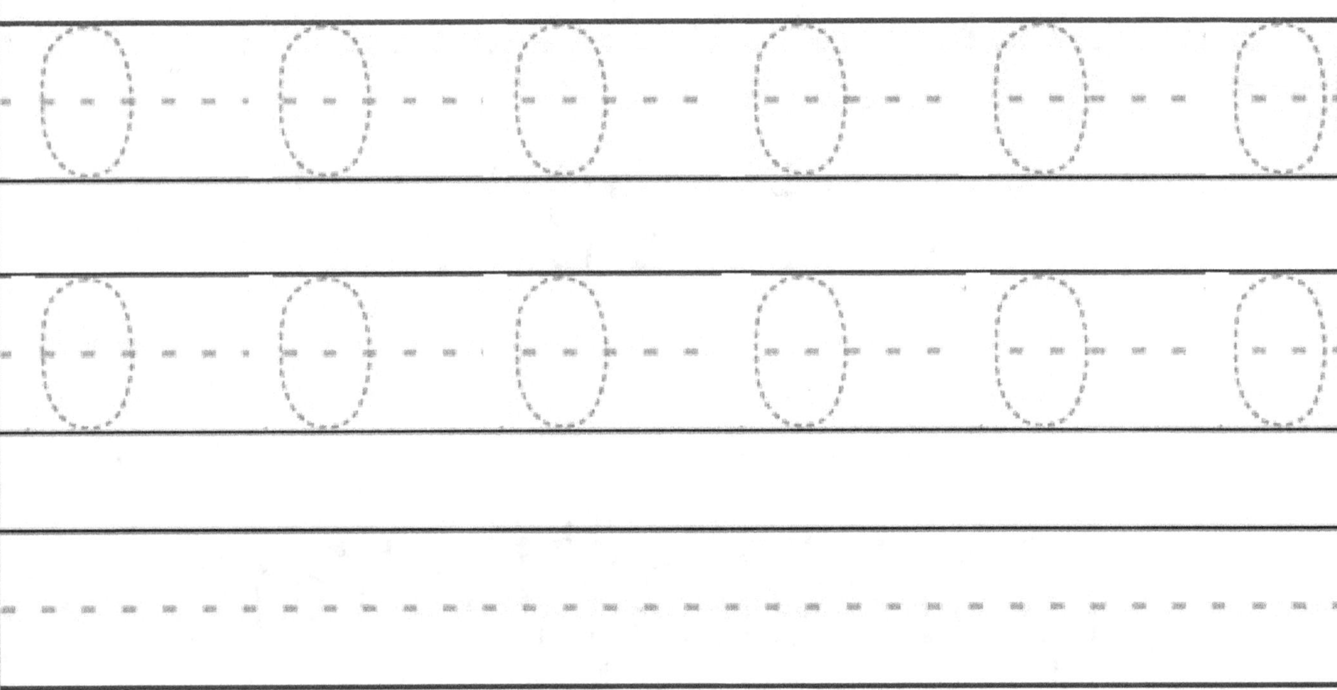

1

one

2

two

3

three

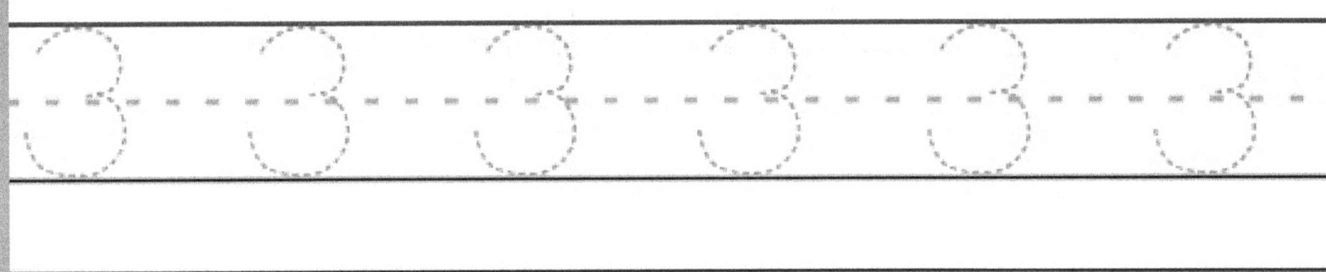

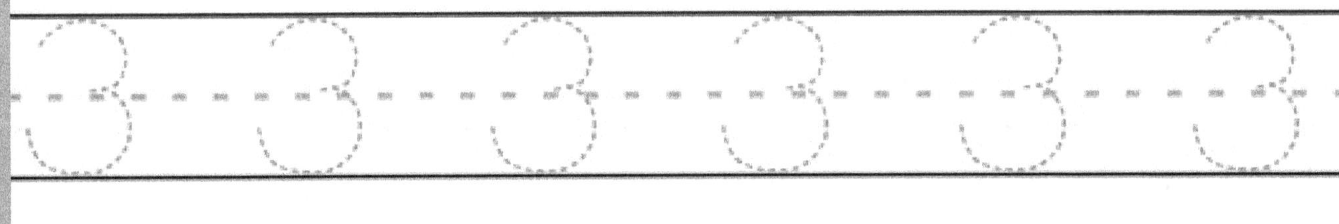

4

four

5

five

6

six

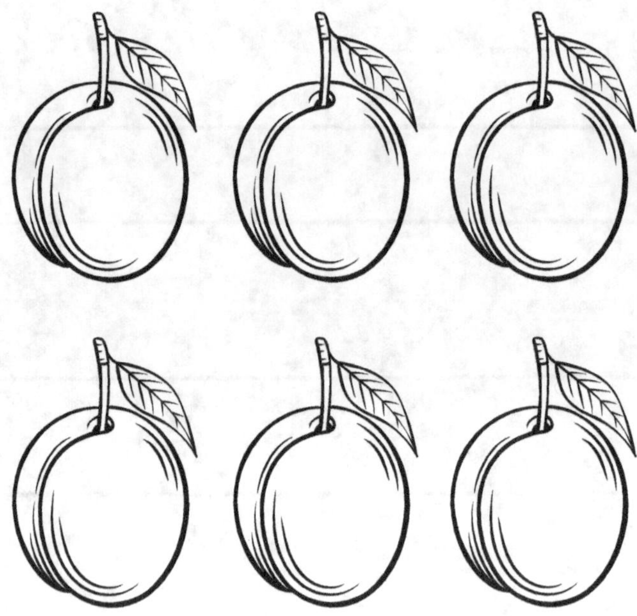

7

seven

8

eight

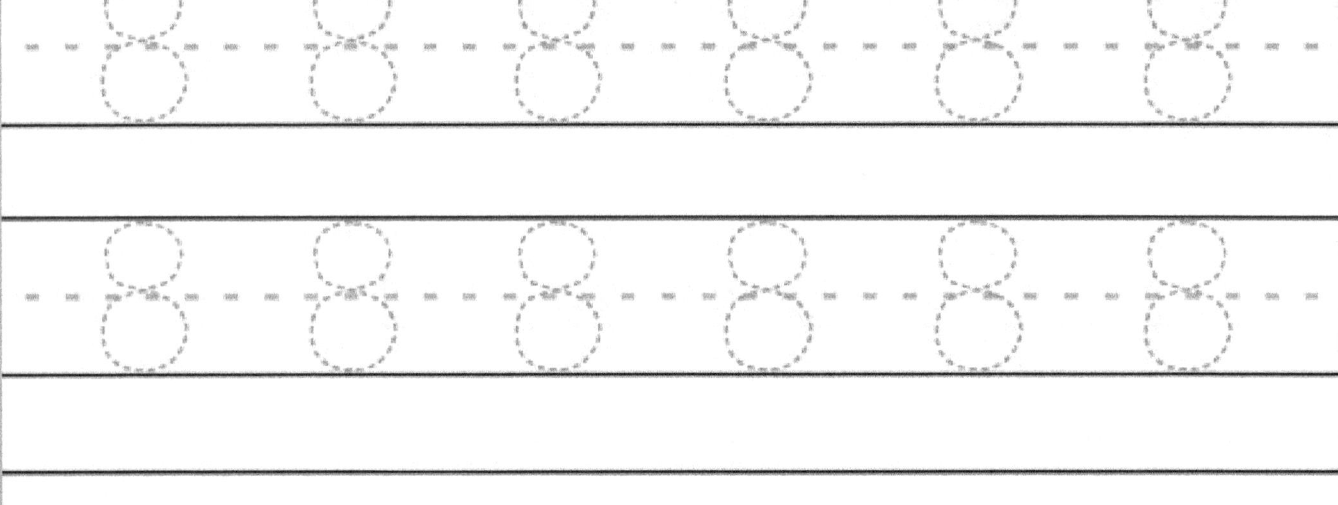

9

nine

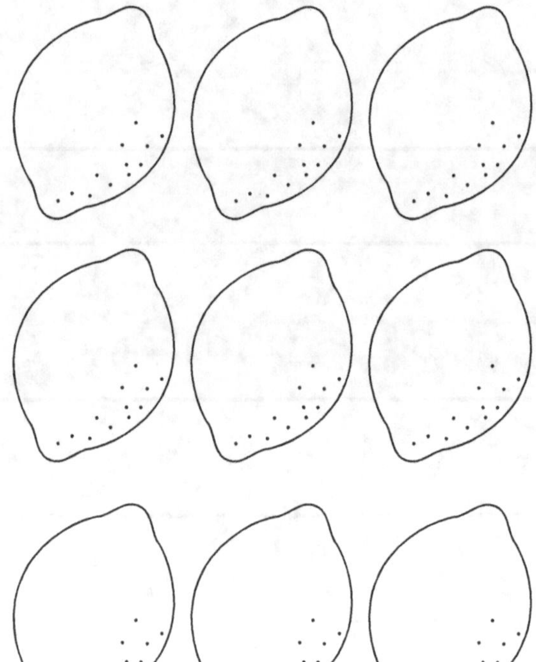

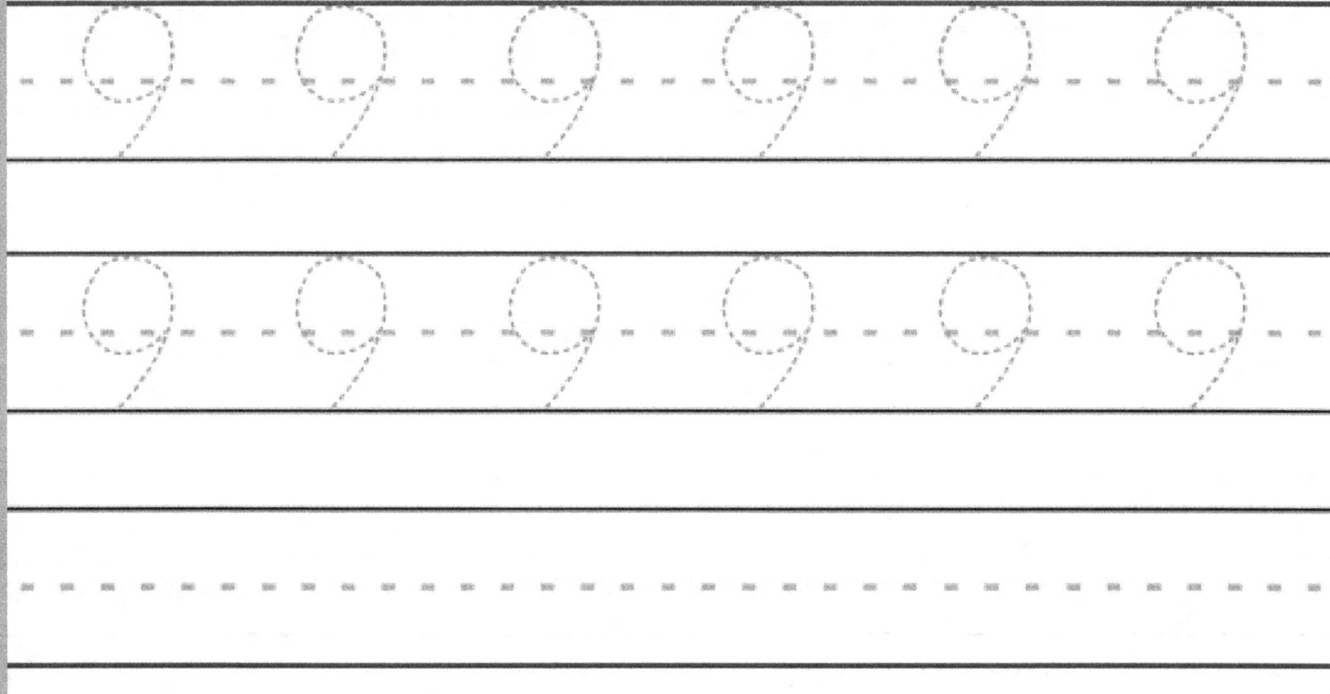

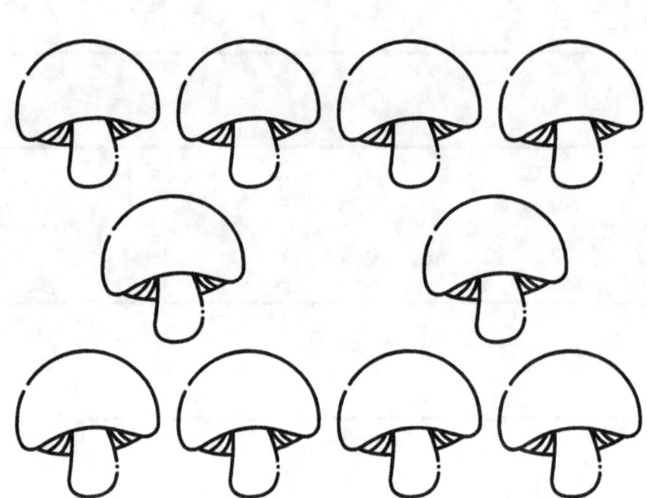

10

ten

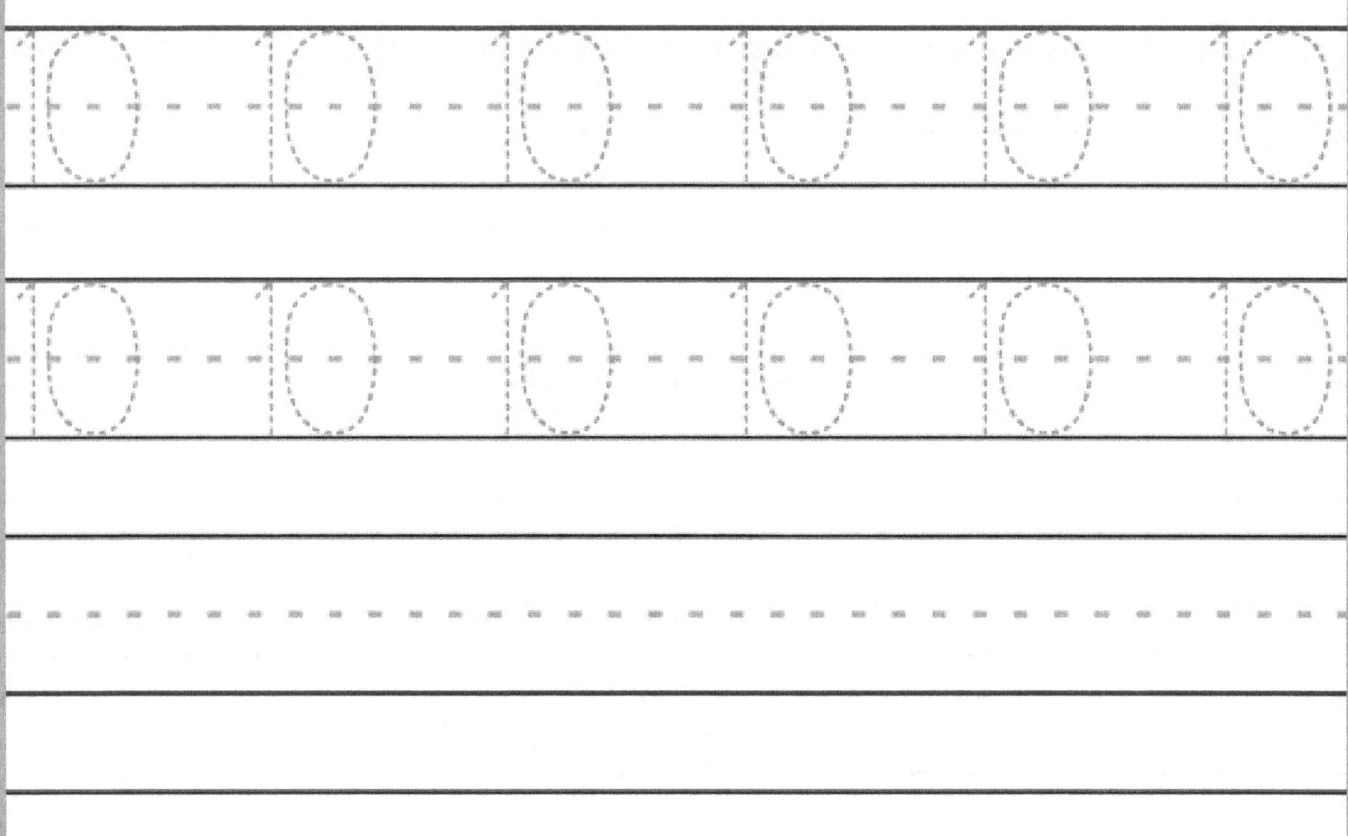

Counting practice

Count the dots and write down the number on the space provided.

7

Count the dots and write down the number on the space provided.

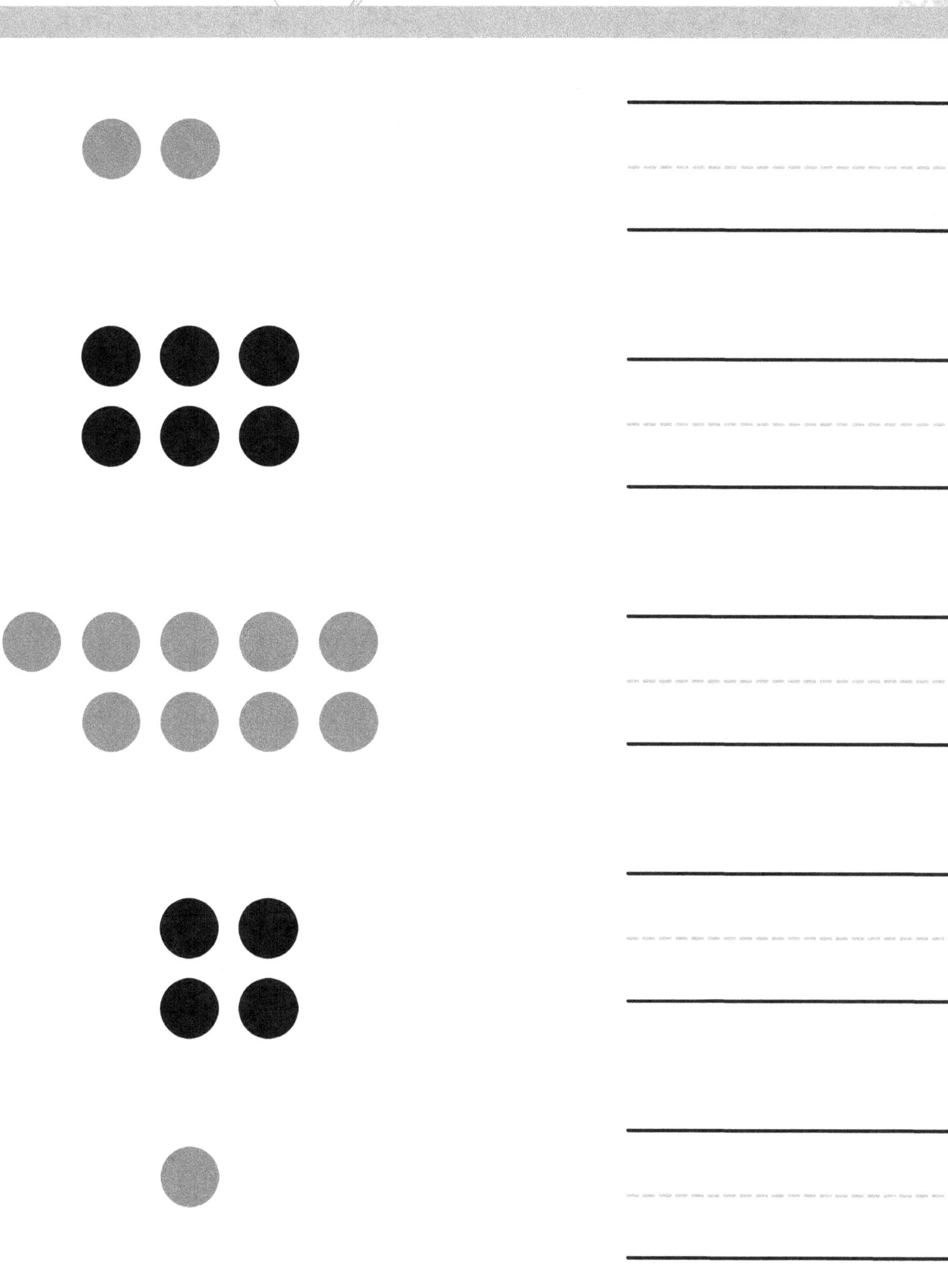

Count the objects in each group and trace a line from the group of objects to the correct number.

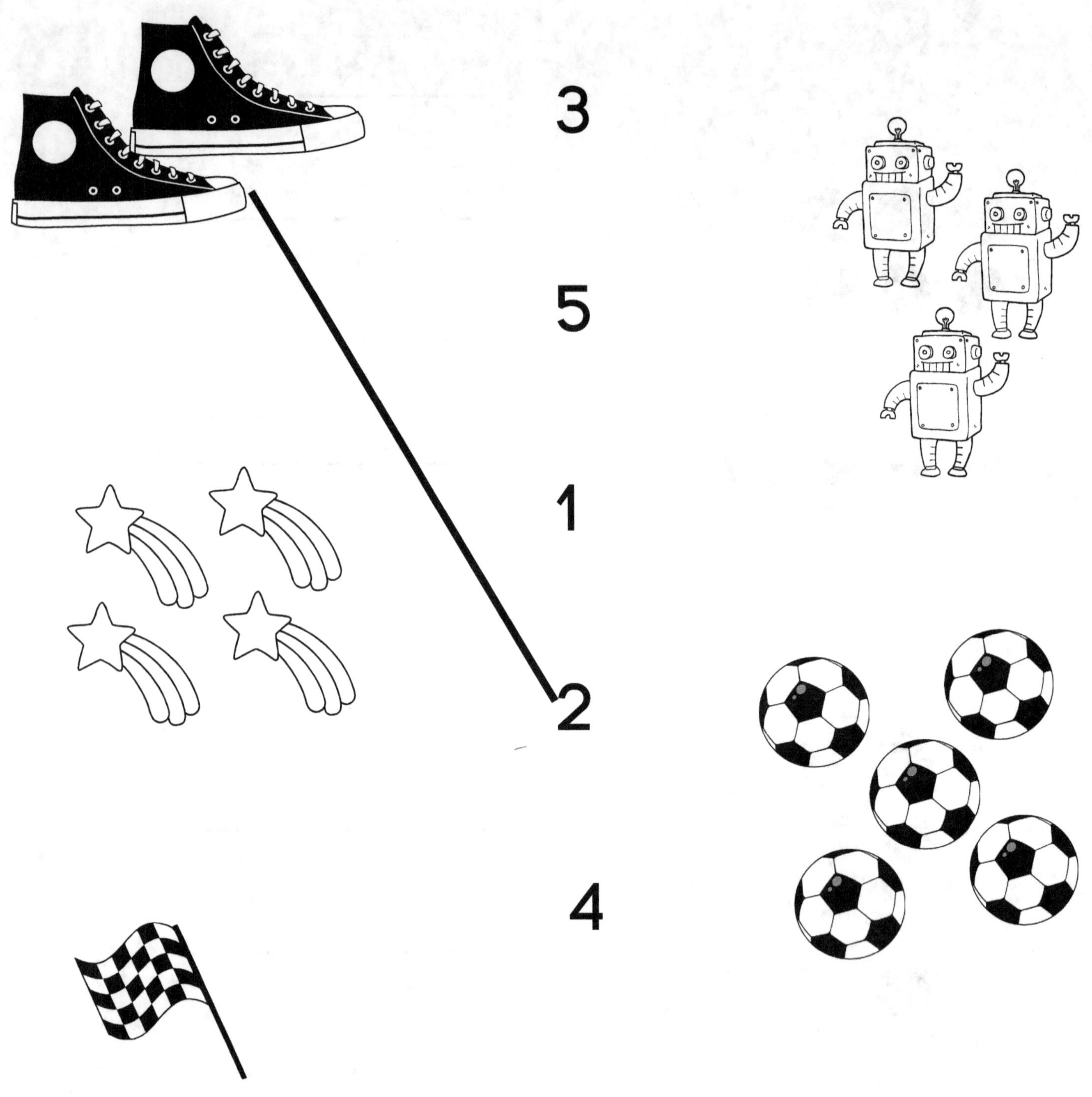

Count the objects in each group and trace a line from the group of objects to the correct number.

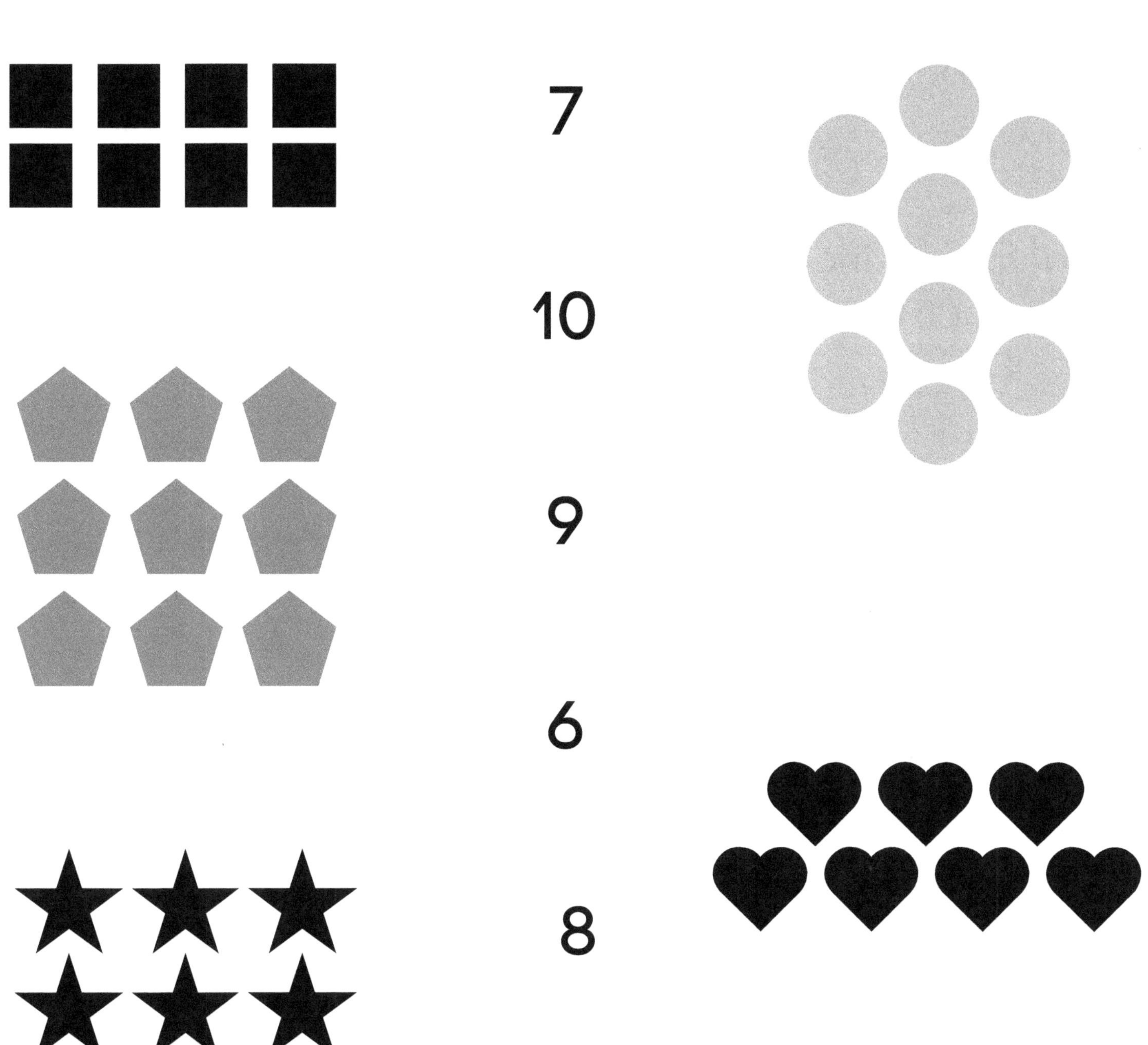

How many people live
in your home?

____

How many windows does
your home have?

____

How many doors does
your home have?

____

How many siblings do
you have?

____

How many pets do
you have?

____

How many chairs do you
see in the dining room?

____

How many kitchens does
your home have?

____

How many phones does
your family have?

____

How many TV's do
you have?

____

How many pencils
do you have?

____

How many flowers do you see?

____

How many chickens do you see?

____

How many cows do you see?

____

How many dogs do you see?

____

How many trees do you see?

____

Count the objects in each group and color the correct number on the right.

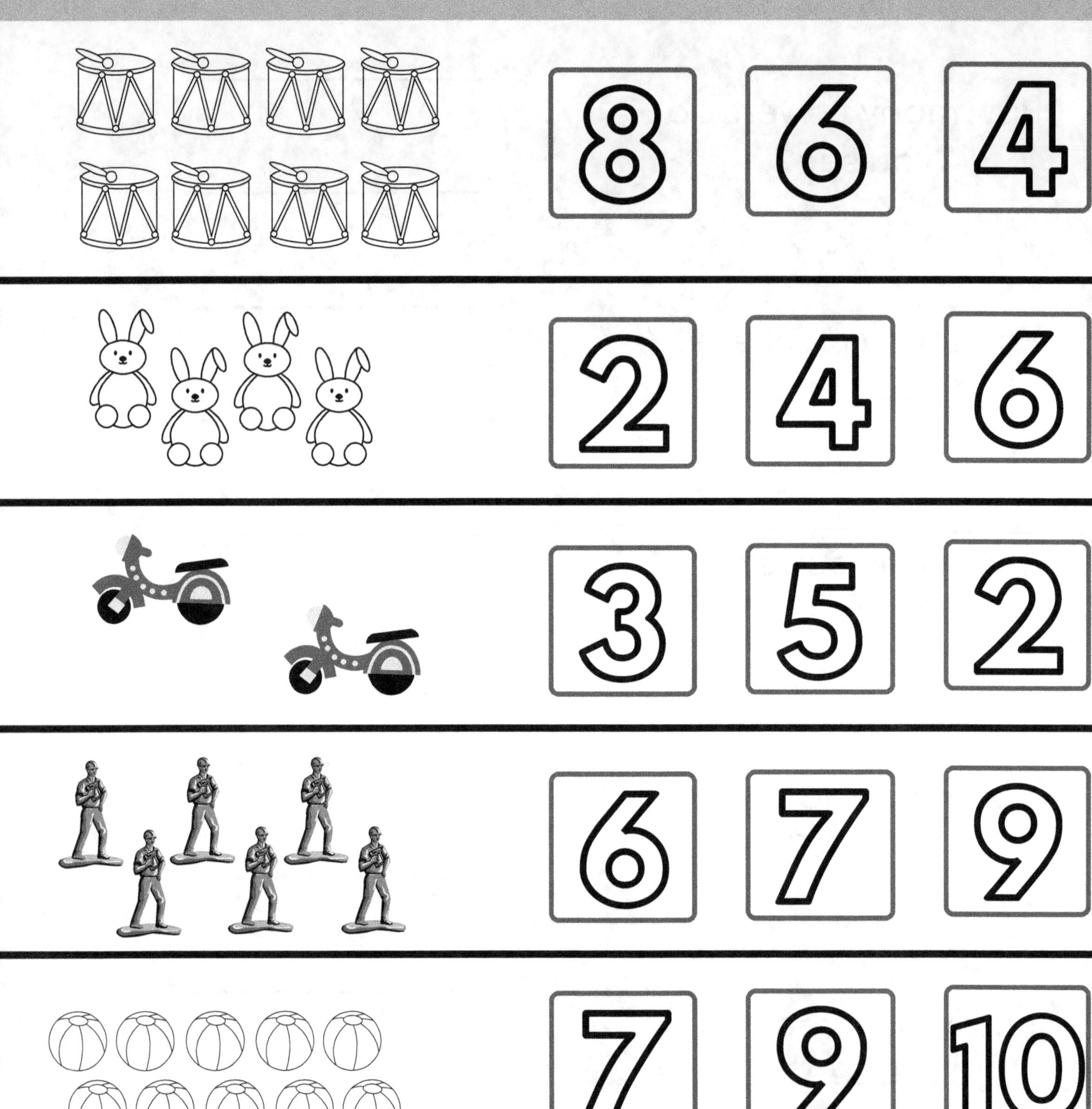

Count the objects in each group and color the correct number on the right.

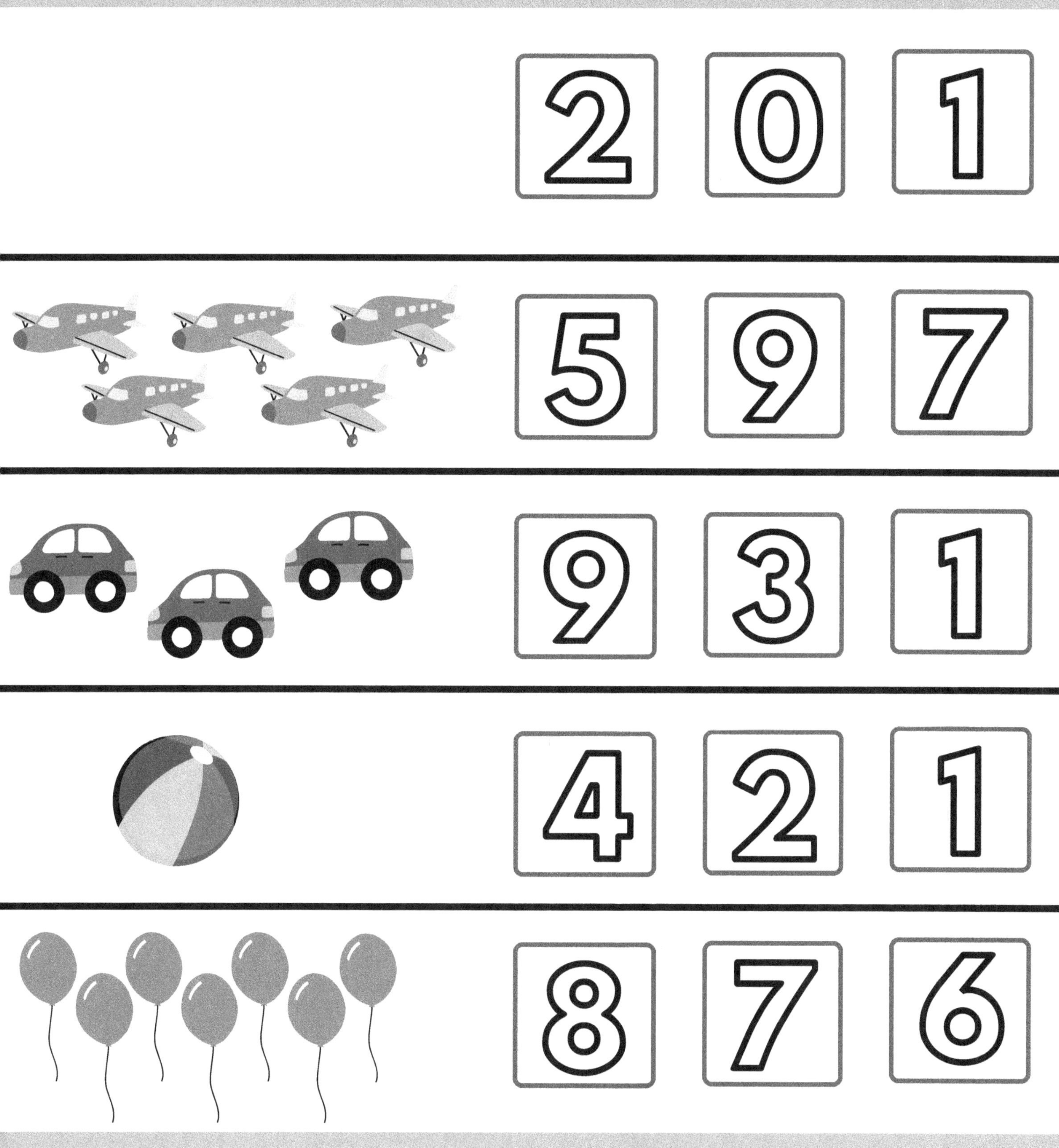

Complete each sequence with the correct number.

1	2	3	4	___
6	7	8	9	___
3	___	5	6	7
5	6	___	8	9
2	3	4	___	6

Complete each sequence with the correct number.

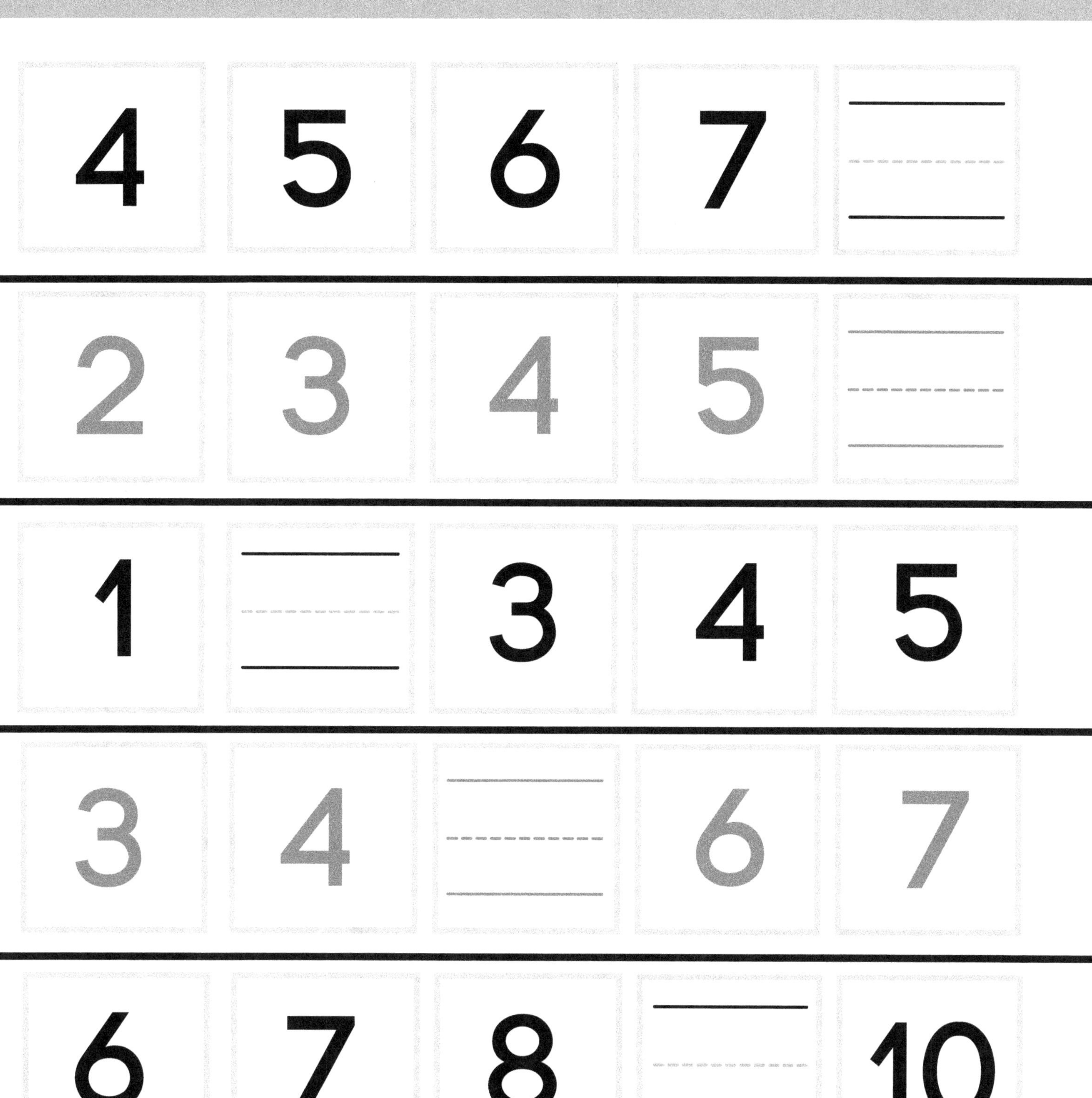

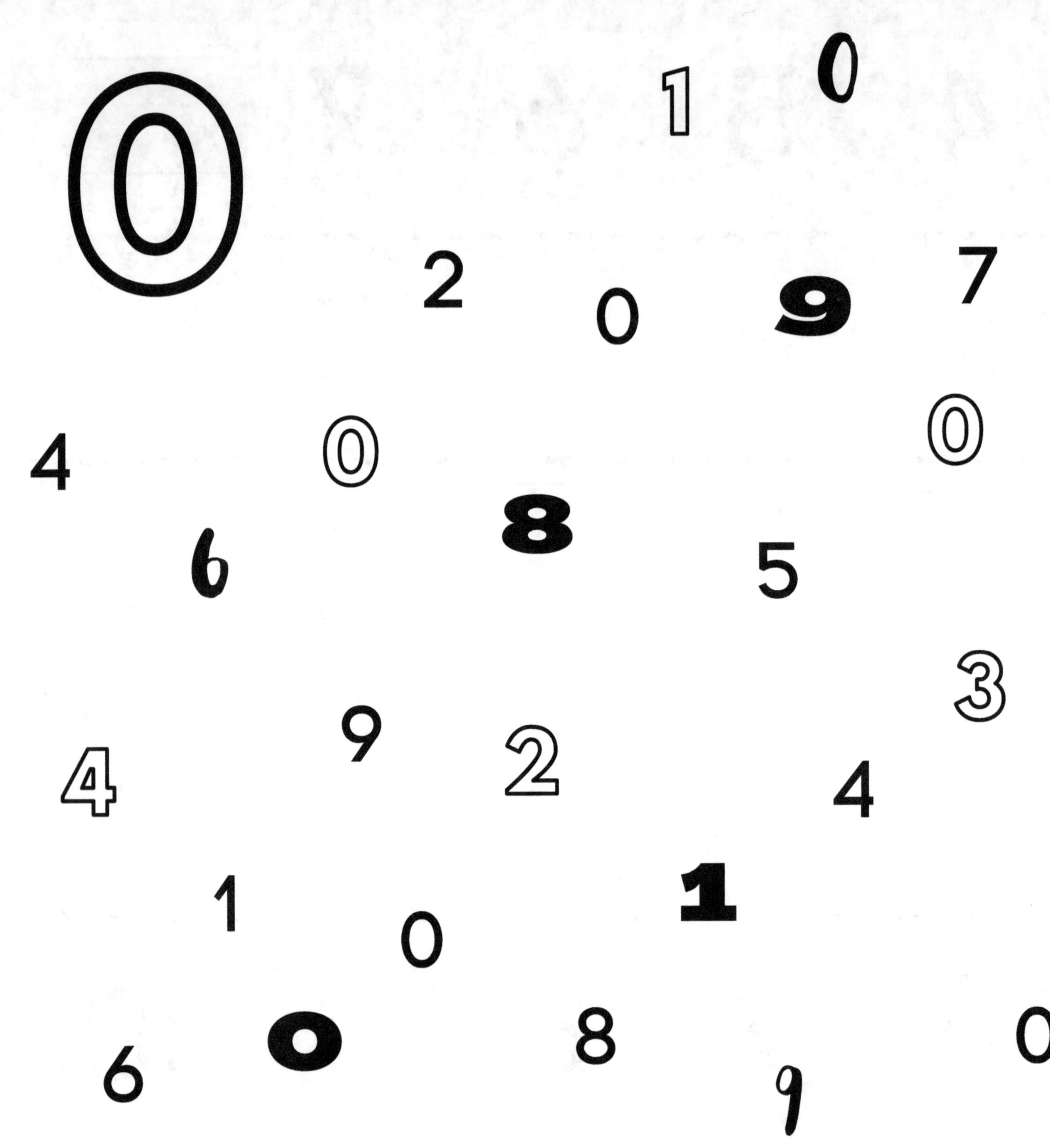

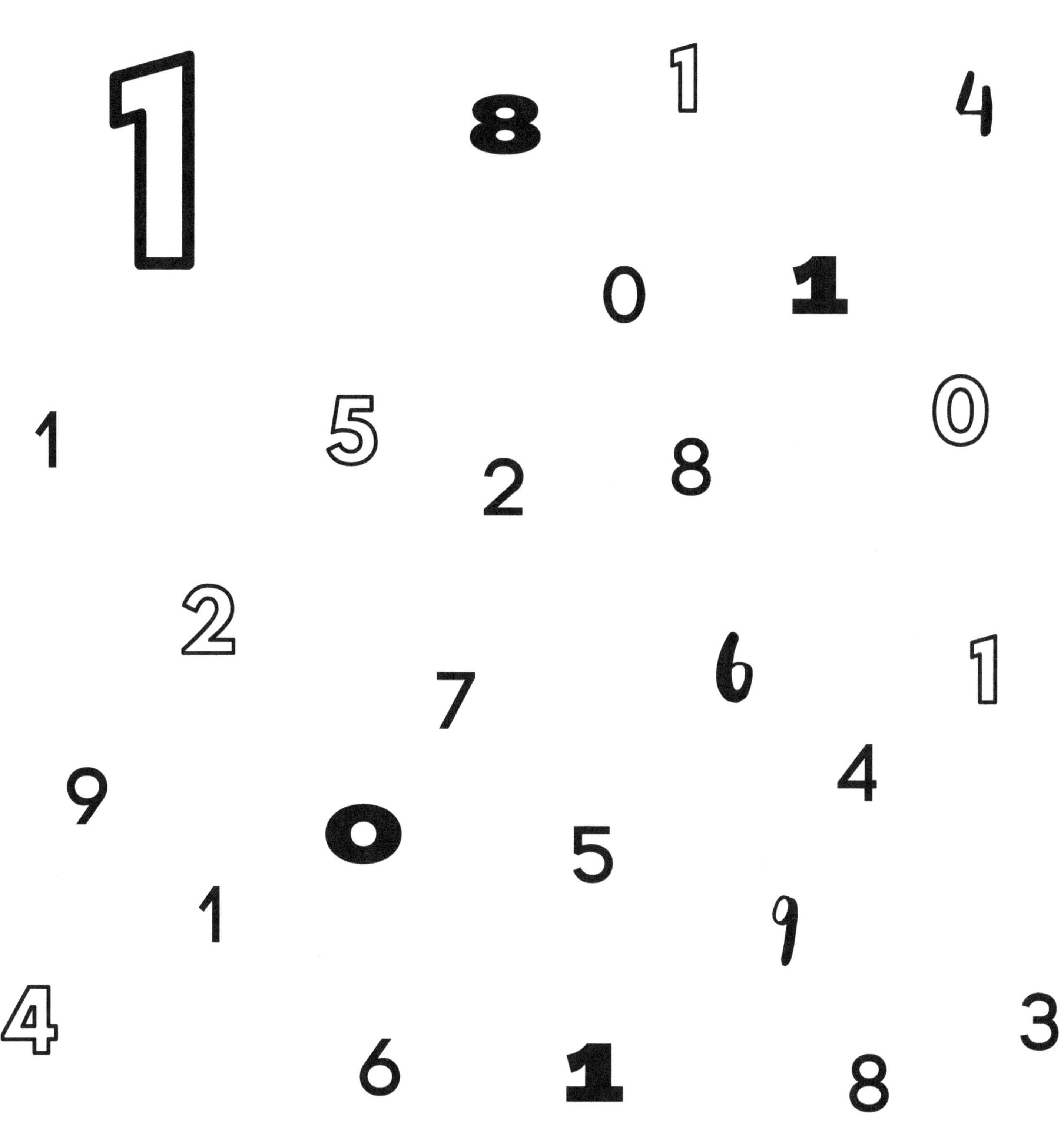

2

3   5

3   2

0

1   4

1   1   8   7

9

2   8   1

7   2

6   1

2   1

7   1   8

4   2

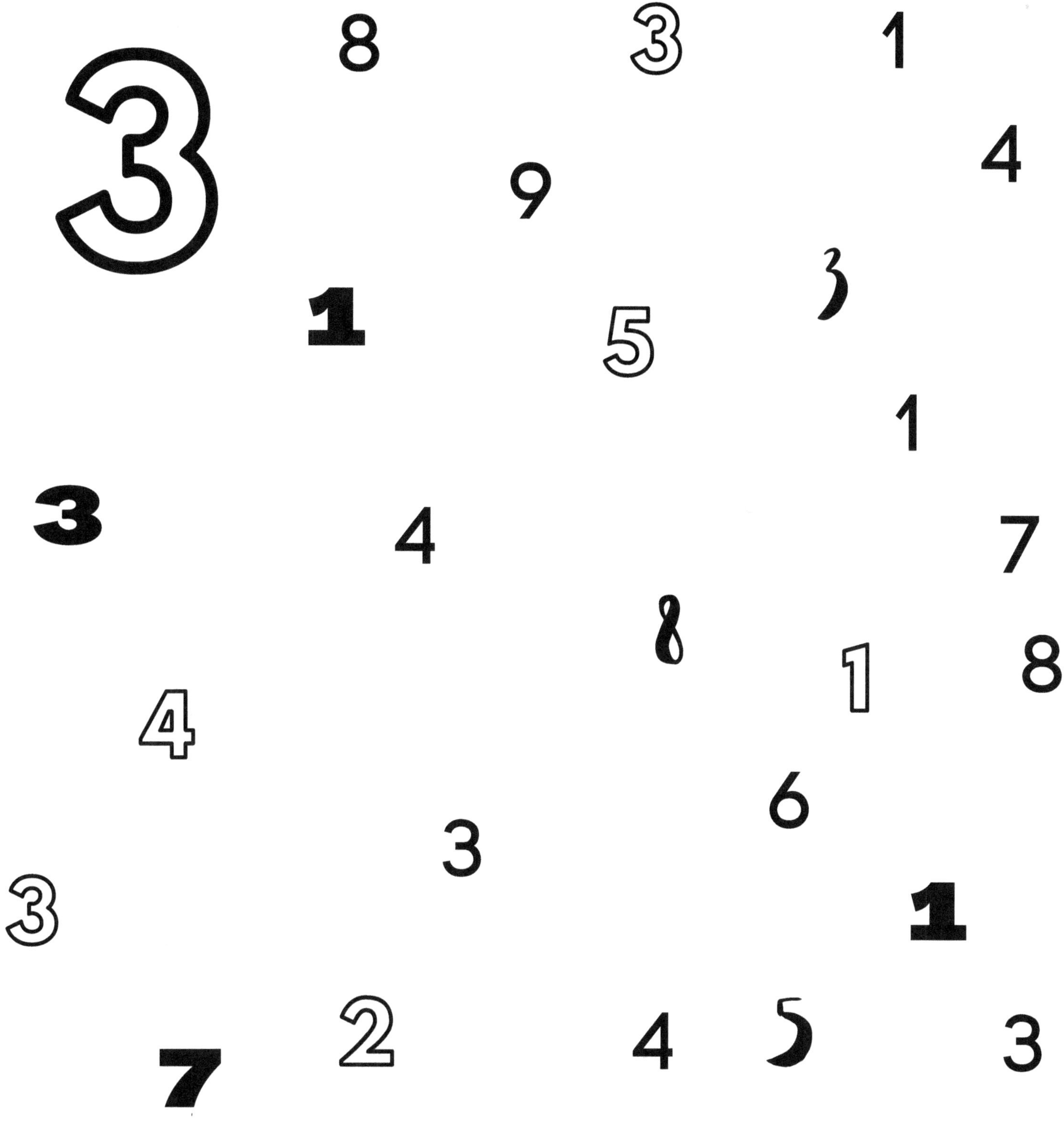

4
1
2
6
2
9
4
1
4
5
10
8
3
4
7
8
4
6
7
3
3
5
4
10
4
9

Circle number 5.

5

1   6   5

10

2

3

1   5

5

7   6

10   3

4

9   5   5   7

4   5

8

8   2

3

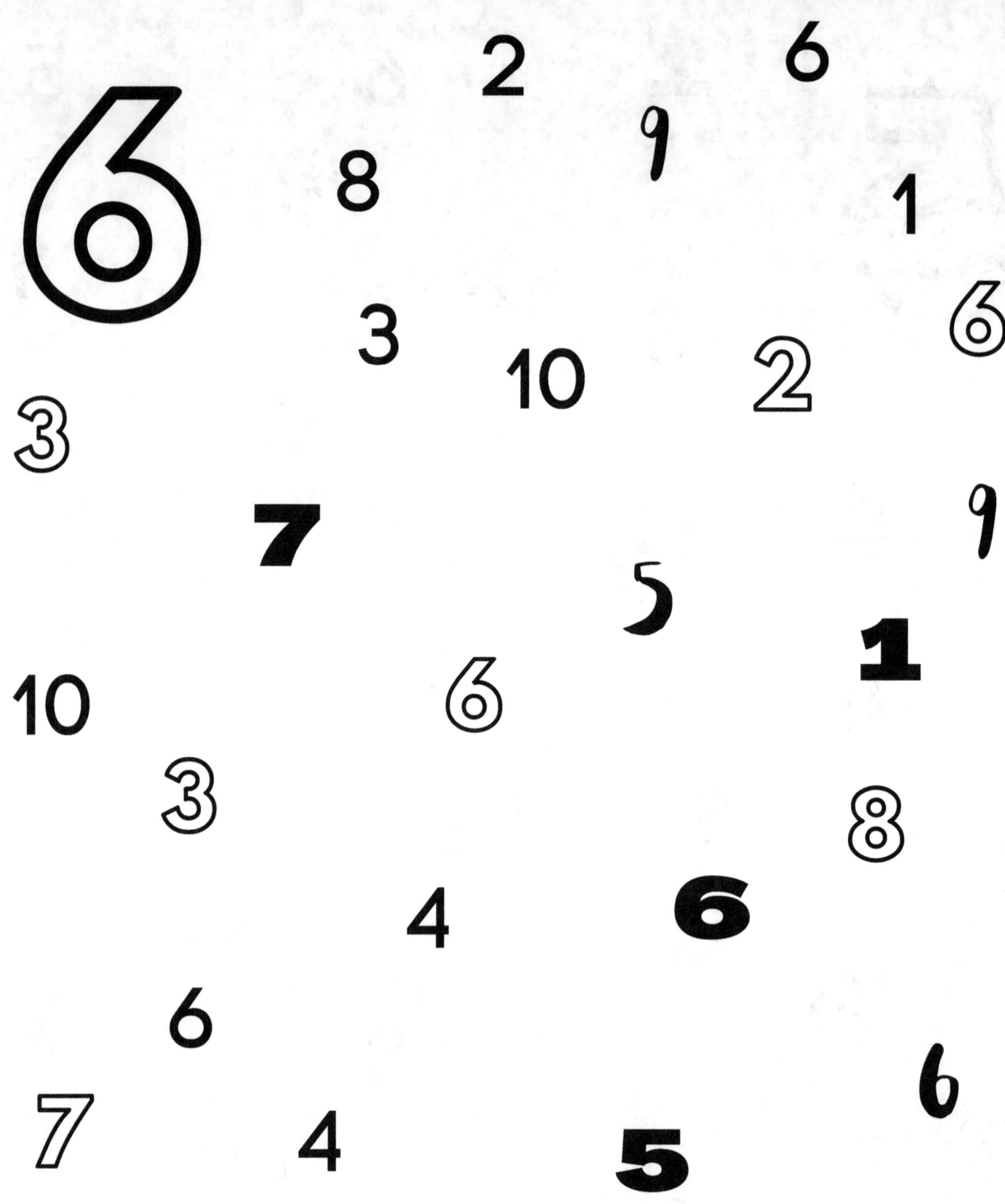

Circle number 7.

Circle number 8.

Circle number 9.

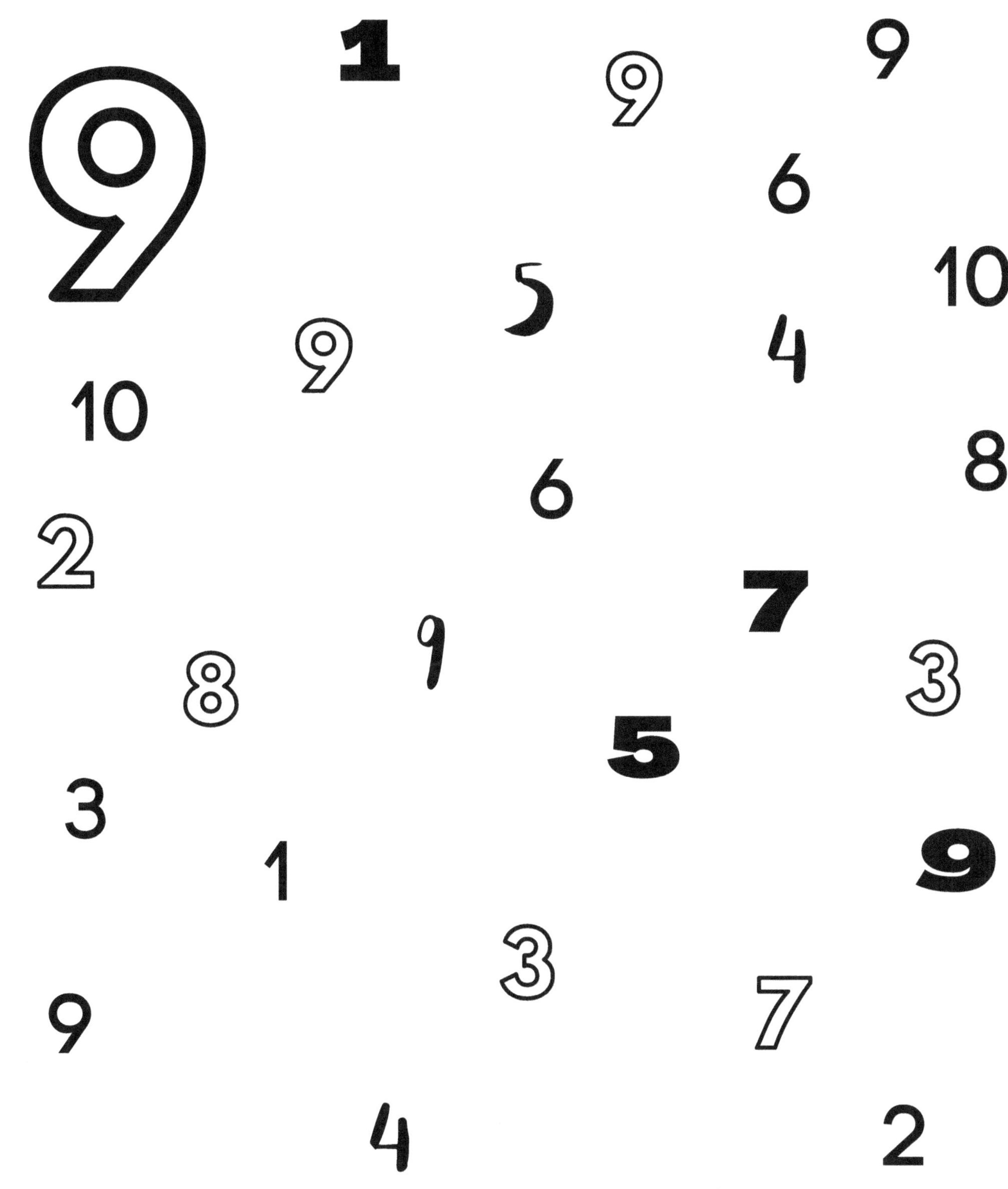

10   2   1

3   10

9

6   5

9   3   10

10

8   7   3

10

5   6   10

8

2   4   1

10   7   4

9 798354 371846